# Turning Humidors with Dick Sing

Schiffer Publishing Ltd

4880 Lower Valley Road, Atglen, PA 19310 USA

# Dedication

To my parents Henry and Leora Sing. To Dad, who taught me everything I know and Mom, to whom I will forever be her little boy.

Copyright © 1998 by Dick Sing
Library of Congress Catalog Card Number: 97-80186

All rights reserved. No part of this work may be reproduced or used in any form or by any means—graphic, electronic, or mechanical, including photocopying or information storage and retrieval systems—without written permission from the copyright holder.

This book is meant only for personal home use and recreation. It is not intended for commercial applications or manufacturing purposes.

Book Design by: Laurie A. Smucker

ISBN: 0-7643-0457-7
Printed inUnited States

Published by Schiffer Publishing Ltd.
4880 Lower Valley Road
Atglen, PA 19310
Phone: (610) 593-1777; Fax: (610) 593-2002
E-mail: schifferbk@aol.com
Please write for a free catalog.
This book may be purchased from the publisher.
Please include $3.95 for shipping.
Try your bookstore first.

We are interested in hearing from authors with book ideas on related subjects.

# Introduction

This humidor is not meant to store large quantities of cigars for a long period of time. It is more of a possession meant to instill a sense of pride by owning something different than Tom, Dick, or Harry (even though Dick made it). This humidor is not meant to be a perfectly air tight, 100 to 500 cigar storage container. It's design is sure to attract the attention of anyone who sees it. You may need to add water to the humidistat more frequently due to the design not having an airtight seal such as in a conventional box style humidor. The round shape and design of the compartments is less conventional than the square or rectangular box. But again, it is not meant to hold large quantities of cigars. I feel the uniqueness of this humidor out weighs its minuses. Stop and think! Isn't the smoking of fine cigars and owning one or more humidors considered a status symbol? Being a woodturner and a cigar aficionado, a round humidor is a natural extension of my craft.

The round humidor does present some challenging problems, but necessity is the mother of invention. The challenges include warpage, a close fitting lid that does not bind, and installing a lining of Spanish Cedar. I do not claim to have the perfect solutions to all of these design problems; however, after serious consideration of the trials and tribulations this project presents, I feel as if I have worked out some of the challenges inherent to the round design. You may expand on my thoughts with completely different solutions. This design may spark your imagination to come up with a totally different concept.

If you are a woodturner who enjoys a challenge, upon completion of this project, you will have the right to say "been there, done that."

I have chosen a staved construction for the body that incorporates small pieces that are not as subject to warpage as larger pieces would be. By using multiple staves, your cutting of the joints must be precise. Not only does this help in making it more stable, it also creates a design that is pleasing to the eye. The lid consists of pie shaped segments to again minimize warpage. We also must use a dried or seasoned wood that has stable properties. Domestic woods such as walnut, cherry, and maple would be good choices. The optimum material would be using stabilized wood, if you want to go to the extra expense to assure stability. There is a down side. Stabilized wood may cause you a problem in machining it. It is more brittle and more likely to chip, especially in woods with a fancier grain pattern.

Humidors require a lining of Spanish Cedar. The reason being, the ability of the cedar to enhance the flavor and smell of the cigars within. It also has an antifungal and insect repelling quality. Installing the cedar lining may sound like a major undertaking, but in reality it is quite simple. This is accomplished by gluing the cedar to the staves prior to cutting the joints. Gluing the cedar to the staves provides additional stability. The same process will be done to the lid and the bottom, making the humidor completely lined.

My choice of glue for the humidor is polyurethane. The main reason being it is waterproof (the humidor is a high humidity environment), it does not creep, and it is also gap filling.

## Reflections -- thoughts, truths, and hearsay.

The optimum temperature and humidity range in a humidor would be temperature - 72 degrees, humidity - 70 percent.

Too much humidity will promote molds.

Spanish Cedar is a fairly stable wood.

Spanish Cedar contains a gum resin that bleeds. It does no harm but is undesirable in appearance.

Quarter sawn wood is the most stable.

Allow wood to normalize before starting. In other words, if it is stored outside in the winter, bring it into your shop and let it normalize to the temperature and humidity before starting the project.

Wood will move during machining due to relieving stresses.

If possible, allow glued assemblies to age or stabalize for a period of time.

Rough turn wood to remove the bulk. Allow it to age and stabalize. This is the optimal approach.

Cherry ages in color naturally. All woods do, but cherry to a greater extent than most due to its sensitivity to ultraviolet rays.

Work proceeds more smoothly with multiple faceplatses. The reason being, we can leave our work pieces attached to maintain concentricity.

Accurate tools are always an asset.

# Turning a Personal Humidor

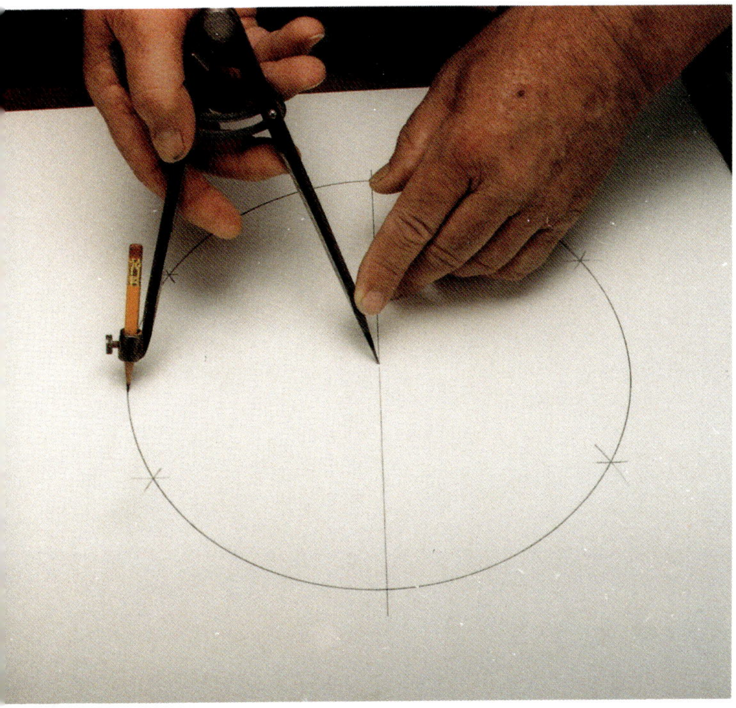

We have to determine the diameter of our humidor. Once we have decided on the diameter, we have to determine the number of staves or segments and the width of them which will establish the rough dimension of the humidor. There are many tables that will establish these dimensions, but for the layman I will work it out mechanically. The outside diameter of mine will be 10 3/4". Draw a circle of this diameter.

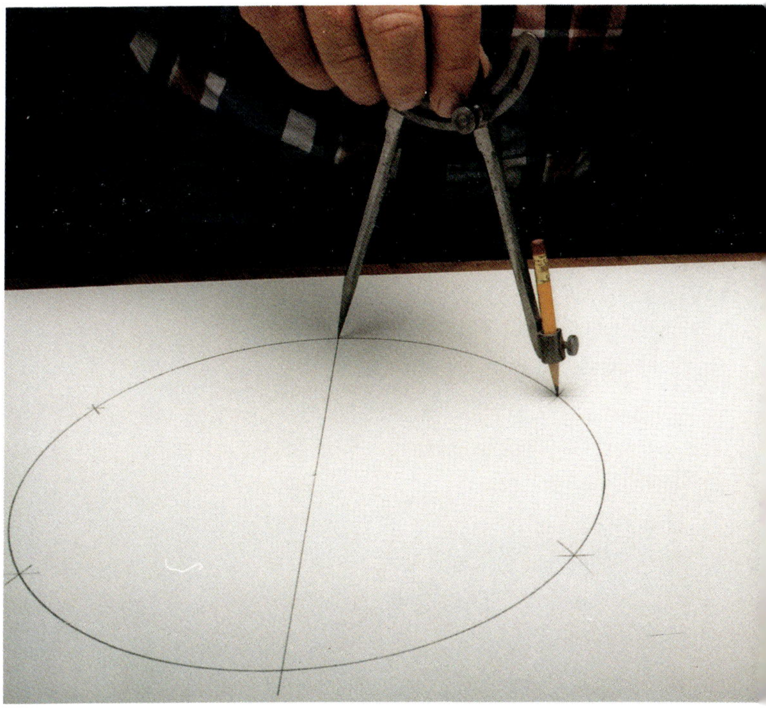

Use the intersection of the diameter and the cross-section line to swing an arc with the compass using the same dimension that was set for the diameter.

Using the center of the circle and a straight edge, divide the circle in half, creating a cross-section line.

Go to the opposite end of the section line and swing the same arc to the diameter.

Swing from the two arcs back to each other. If you are precise in your measurements and pickups, all of the points will agree. Do the same to the opposite half circle.

By connecting the center of the circle and opposing arcs, we have established six segments.

Using the compass, divide each segment in half.

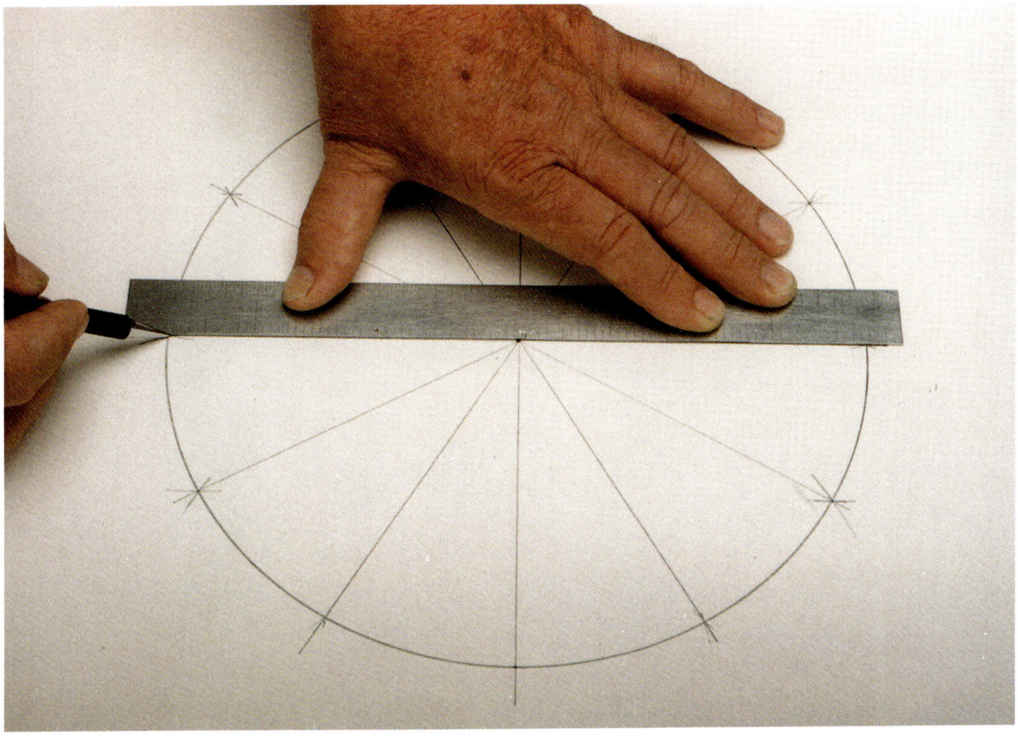

Again, using those intersecting arcs and the center of the circle, connect the lines. We now have twelve segments. I have chosen twelve segments and staves. The reason I have done this is that with smaller pieces there will be less warpage or movement. Using this layout, we can now establish the width of our staves.

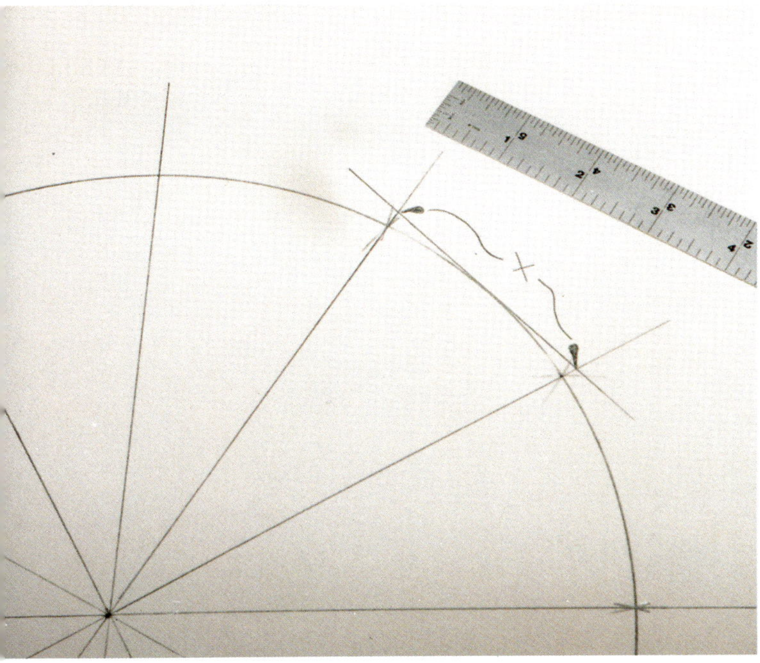

Extend the two segment lines past the diameter of our drawn circle. Using a line across the diameter, intersect the two segment lines. This line must intersect the two segment lines at equal distance to the center of the circle. The "X" dimension will be the width of the segment needed to produce this diameter. If your humidor is chosen to be a different diameter, adjust your measurements to your needs.

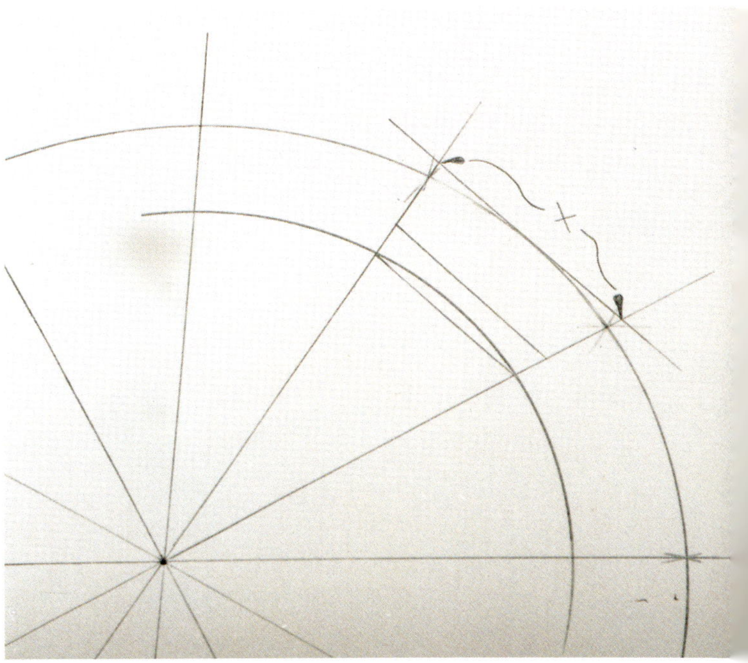

We can draw in the thickness of the wood and visually see what the inside diameter will be. This is done by using whatever thickness lumber you desire.

Twelve segments will equal a saw setting of fifteen degrees. If you choose to use a different number of segments, most woodworking reference books will give you the segments and angles.

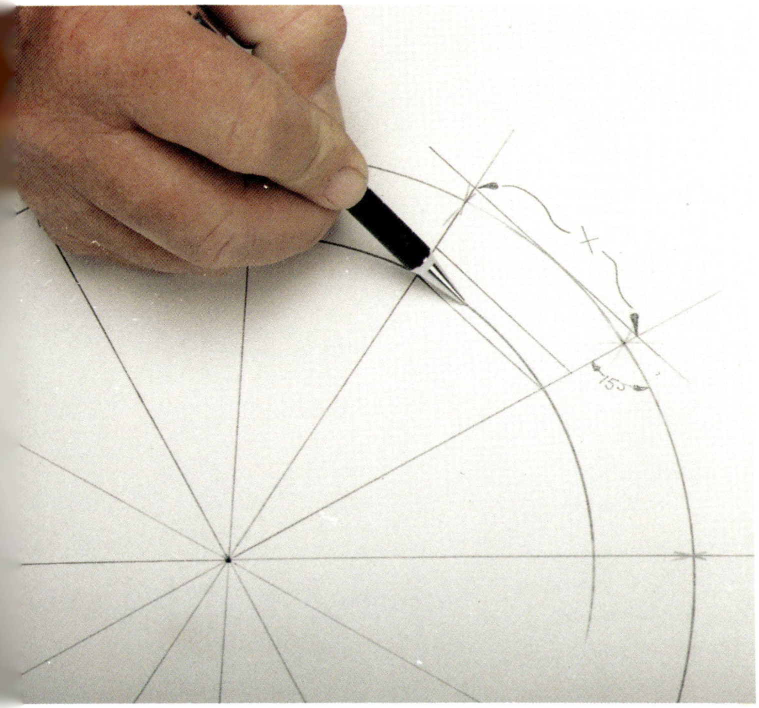

This line represents the minimum cleaned up inside diameter.

Let's start construction. For this humidor, I will be using cherry with a liner of Spanish Cedar. The initial reaction in lining a round object sounds very difficult. In reality, by gluing the cedar to the cherry prior to cutting our segments, the task becomes simplicity itself. To start with, both the cherry and the cedar should be of their own uniform thickness. The cherry's thickness will depend on your outside design. Mine is approximately 7/8" thick. To arrive at a cedar lining that is approximately 1/4" finished thickness, I am using 7/16 cedar stock. These dimensions were determined by plotting on the initial layout. Again, I stress, to make this work, the laminations must be of uniform thickness.

I am going to take the liberty to assume your abilities to cut angles on your table saw, radial arm, or whatever means is at your disposal. I use both radial and table saws.

After cutting, you can see the lamination between the cherry and the cedar. Make sure the cedar is on the inside before you cut. Look twice, measure twice, cut once. The length of the staves is determined by the height that you intend your humidor to be. I have chosen approximately 5" which will give me approximately 4" inside height.

Preassemble the staves and align opposing grains wherever possible. This will help to make it more stable and also to make the outside more eye-appealing. If the joints do not align, now is the time to correct them. We do not want glue to have to fill gaps.

To clamp these, I use hose clamps. To obtain the diameter clamps needed, I take two or more or the same kind and connect them. Also, with more than one clamp you can draw from opposing sides which helps to clamp uniformly around the body.

To facilitate reassembly during the gluing process, number the staves.

First, be sure to cover your work surface with wax paper as glue will not stick to it. This will also help to keep your work surface clean. My choice of glue for this application is polyurethane. The reason being, it is water proof, gap filling, and does not creep. You may also want to use latex gloves to keep the glue from your hands. If the glue dries on your hands, it stains them. The only way to remove the stains once dried is to wear them off. If you do get glue on your hands, denatured alcohol, mineral spirits, or most solvents will remove it before drying. I prefer to use a flexible bladed putty knife to spread the glue uniformly. A thin uniform coat will do because of the expanding qualities.

Put your numbered sections together and apply the band clamps. As you tighten them, tap the segments into proper position vertically and horizontally.

Using a solvent, clean off the ends of the staves as we need to set up on these. This will remove any surface irregularities created by the expanding glue.

The lid. When making our lid segments, a simple way to rough cut them is to use a board slightly wider than the width of your segment. Start by laying out a 15 degree line, moving up the length of your segment and putting another parallel 15 degree line and connecting them corner-to-corner. This will save you lumber and allow you to oppose grain.

Cut your twelve segments minus the cedar which will be added later. Dry fit and number them in preparation for gluing. Again, pay attention to your grain direction as warpage is even more critical in the lid.

Spread the glue in a thin, uniform coat on all twelve segments as before.

Tighten down the segments and wipe off the surface with the solvent for later ease in set up.

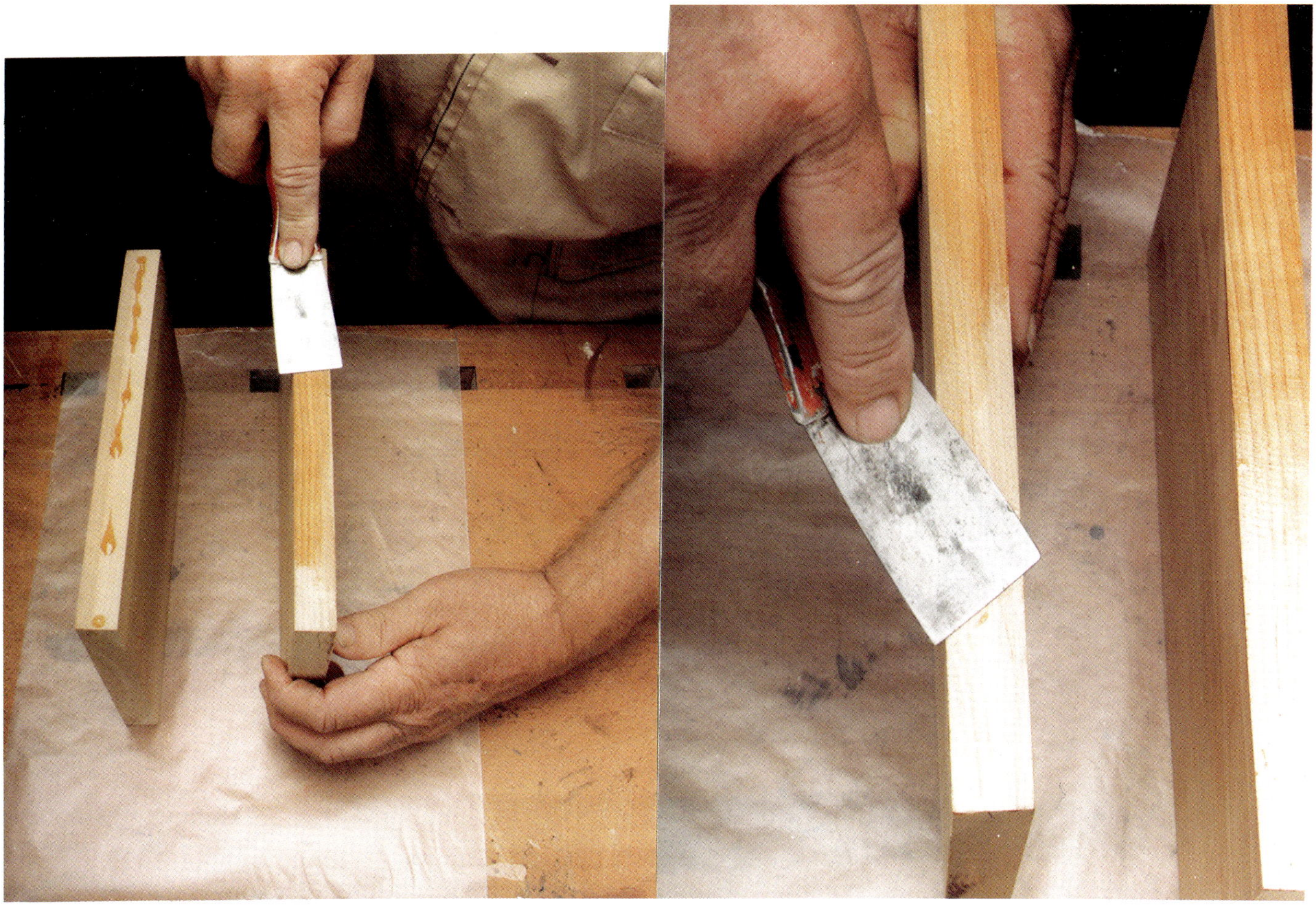

The bottom, minus the cedar lining. Select two pieces of cherry large enough to cover the bottom. Join the two pieces together, paying attention to opposing grain. Again, the same as the top, the cedar lining will be applied later.

Apply a uniform coat of glue. Clamp, making sure both pieces are matched. Wipe off excess glue from the joint line.

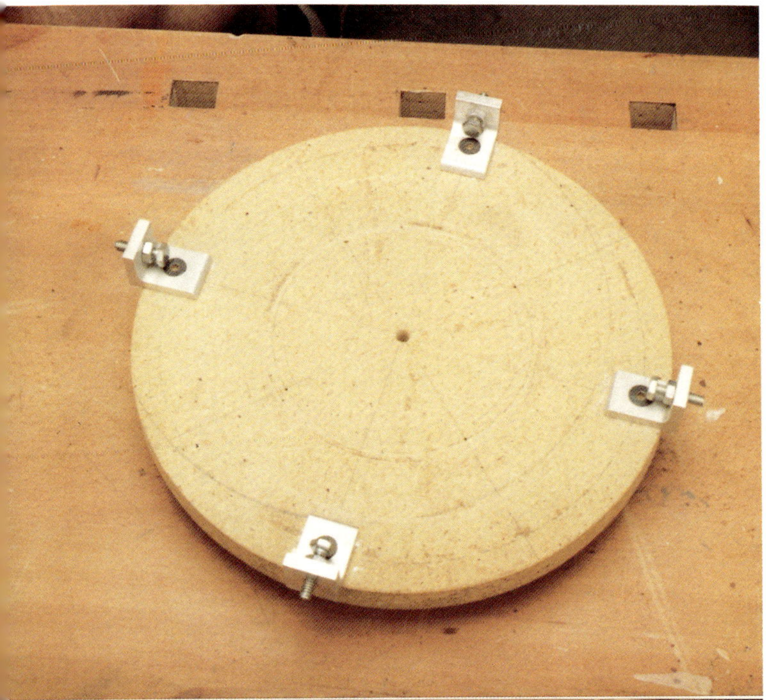

Take the lid and determine the top which is probably the side you have numbered. Clean off any glue or obstructions and flatten out the surface.

Attach a scrap block to a face plate and face off the surface. This will become the surface the lid will be attached to.

When it comes to turning the barrel of the humidor, I have made a jig which will hold my staves so that I can turn them. This consists of a piece of 3/4" particle board mounted on a faceplate. I have also mounted four pieces of sturdy aluminum angle spaced at quarter intervals. These are held on with bolts and T nuts. On the opposite arm of the angle I have drilled and tapped four 3/8 holes hex head bolts. These bolts also have a lock nut. This allows me to adjust the position of the barrel until it is concentric to the axis of the lathe.

Apply several pieces of double faced tape to the scrap block and screw it onto the headstock.

On the bottom side of the lid, take an awl and find the center (the intersection of the points of the segments) and mark it. This point will be used to center the lid to the faceplate.

Put the point from the live center into our awl mark and carefully slide the tailstock and work piece up to the face plate. Before applying pressure, rotate the lathe and let the stock seek its center on the faceplate. If all is well, apply pressure from the tailstock.

Clean up the diameter using a 3/8 spindle gouge. Remember you are working on end grain with joint lines. Cut with caution.

Using a sheer cut, make sure that it is flat and uniform as this will become the surface that we will glue the cedar to.

Back off the tailstock. Clean up the face lightly.

Remove the lid from the faceplate. The easy way to get double face tape to separate is to apply continuous pressure and just hold it. This will allow the tape to relax and separate. Mark the center on the top of the piece and relocate back on the faceplate in the same manner as before. The surface we are working on now will be the top.

Face off the top of the lid. This will now become our holding surface.

I like using a screw chuck so I will be working the rest of the lid off this type of chuck. The screw in my chuck necessitates a 1/4" pilot hole. When bringing up the tailstock, allow the drill to seek its center before tightening down the tailstock. This will assure an accurately centered hole of the proper size.

Remove from the faceplate. Cut enough 1/4" Spanish Cedar to cover the top of the lid. The reason I am doing it in this stage is that I feel larger pieces glued to the segments will give added support and stability. Glue and clamp.

Affix the blank to the faceplate with the double face tape, centering it with the tailstock and center hole as previously done. Clean up the diameter.

Take the pieces we glued together for the bottom and lay out the diameter. Go to the bandsaw and cut out the bottom blank.

Face off the surface as we will be gluing on a scrap block to attach a faceplate.

Remove from the taped faceplate.Glue on another faceplate with the faced off scrap block. The barrel will be turned on this faceplate and scrap block throughout the remainder of the project. I am using Titebond II as it is much faster drying and it will be cut off later anyway. The cherry used for the bottom is thicker than necessary but will not end up being this heavy. The reason I am using such heavy stock is to help dampen vibration during the turning of the humidor, inside and out. Set aside to dry.

Put the barrel section into our jig. Using the four bolts, center it to the axis of the lathe. Tighten lightly and lightly scribe with an awl such as at the joint between the cherry and cedar. If all twelve points that you choose are equal, it is centered. Tighten up the bolts reasonably so as not to split or push it out of round and tighten the jamb nuts.

For safety's sake (so that your wife can sleep at night), and to the dismay of your dentist, make sure that the barrel is tight against the foot of the aluminum angle and use duct tape to assure it is staying there. This also will help to dampen vibration. We are ready to turn.

Face off the bottom of the barrel. At this point we need it to be flat. We are going to be using it as a work surface to cut a rabbet for the bottom.

Start turning the inside of the barrel. The inside surface should be turned round with the cedar thickness left to a minimum of approximately 1/4". Take care in cutting the cedar as this will be close to a finished surface. It will be difficult to reach from the top when we reverse the piece. Use a sheer cut on the inside to make it as smooth as possible.

Bridging across the bottom with a straight edge, use a square to check that the inside wall is 90 degrees to that surface. We have to take pains here because we will be matching the other half coming from the opposite direction when we reverse our work piece.

Cleaning up the wall of the rabbet.

Starting to cut our rabbet so we can recess the bottom into the barrel. This way, the bottom will not be showing, only the staves will be visible from the outside.

Using a square, check to make sure that the rabbet is 90 degrees to the base. The rabbet is approximately 7/16" deep. If we are to err, make sure that we err in the direction of the corner so that our matching surfaces will be a tight fit. Also, if you notice, I have turned away all of the cedar except on the top of the rabbet so as I will be gluing to the cherry. This should help assure a more stable joint. Interior dimension of the rabbet is not critical as the bottom must be fitted to it. Do not remove from the jig at this time.

Back to the bottom. When the glue is dried between the scrap block and the bottom, remount on the spindle and face off the surface. Glue and clamp the cedar, laying the cedar 90 degrees to the grain pattern of the cherry. Allow to dry.

When the glue is dry, remount the assembly on the lathe and clean up the diameter.

Take care not to reduce the diameter smaller than the rabbet in our barrel as this needs to be a close fit.

Clean up the face of the cedar. Take care while working on this surface as it will be the base inside the humidor.

As we get to our final sanding, you can use a sanding block. This will help maintain a flat surface. Stop the lathe and finish sanding with the grain to eliminate scratch marks. At this point our sanding should be finished on the bottom.

When you have cleaned up the surface, leaving approximately 3/16" to 1/4" of cedar - or whatever you desire - it's time to sand using a power sander and drill motor. Select the appropriate grit for the surface you have left and do your best to maintain a flat surface as this will be the bottom of the humidor.

It is now time to fit the bottom to the barrel. Check your dimensions, the inside of the barrel rabbet, and the outside of the bottom.

Work carefully and maintain a 90 degree corner as we want this to be a good glue joint to our rabbet.

Remove the assembly from the lathe and check often to make sure we do not remove too much stock. Return the assembly to the lathe and lightly remove stock until our fit has been achieved.

Keep fitting until our piece presses in with a light whoosh of air. This will allow a coat of glue yet will align everything perfectly.

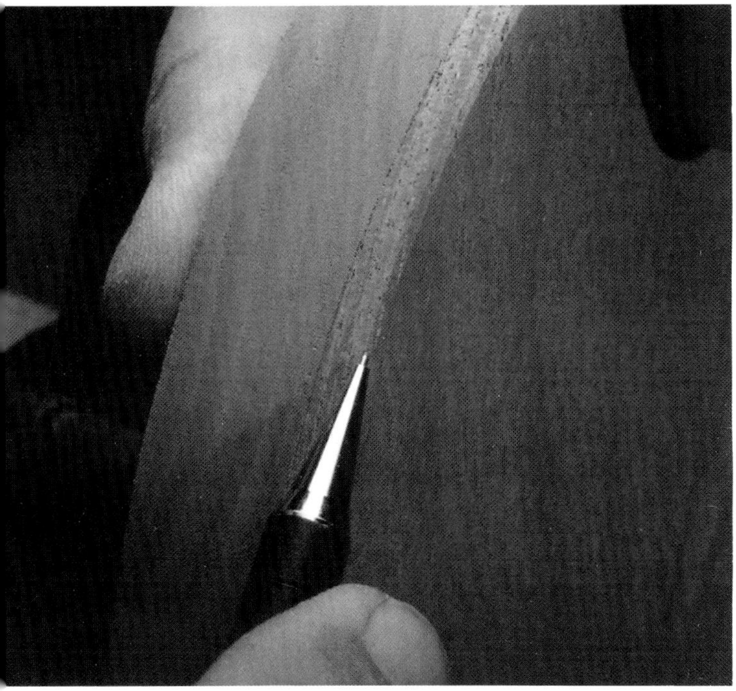

Put the bottom back in the lathe and break the corner. This will assure there is no interference with the inside corner of the rabbet, ensuring a close matching fit. We are finished turning the bottom but do not remove the faceplate from the scrap wood as this will be our working surface for the remainder of the humidor.

Test fit the bottom to the barrel and make sure our joint is right. After we glue it, it's too late!

Put the jig and barrel assembly back in the lathe and sand the bottom inside of the cedar. This will become difficult to reach later. Use care to maintain a sharp corner where it matches the bottom. We can now remove it from the jig.

When we apply glue, try to stay away from the edge so there is no squeeze-out into the inside of the humidor. Why clean up something unnecessary if we take steps to prevent it?

Back to our lid. Now that the glue and our cedar has cured, it's time to remount it. I am going to use a screw chuck. If you do, make sure the length of the screw does not protrude into the cedar and damage it. If your screw is too long, you could make a spacer to reduce the length of the screw. Other than a screw chuck, you could also use a scrap block with glue or double face tape. If you've already come this far, I am sure you will be inventive enough to use one or the other.

Mount the chuck and lid on the spindle and clean up the diameter and the face.

Put a live center and 60 degree tip in the tailstock. Use a piece of scrap or the lid across the top of the barrel. I have used the lid and located off the screw hole. Apply light pressure to the tailstock. This will help eliminate vibrations while we turn the outside of our barrel. It also will give us an idea of how it will look with the lid.

We have removed the screw chuck and the lid from the lathe. Remount the barrel assembly and clean up the top surface with a 3/8 spindle gouge.

Cleaning up the barrel with a gouge.

Once cleaned up, start paying attention to the shape that you want. When it comes down to form, it's not what you take off, but what you have left. You can also reduce the diameter of the lid at the same time, making sure that you do not reduce it past its usable diameter.

Start cleaning up the inside of the humidor.

Blend back to our sanded surface.

You can power sand the surface with the appropriate grit, I used 150 grit.

Block sanding will make a more uniform wall.

The inside is sanded. We are now ready to start making our recess for the lid.

Using a 3/8 spindle gouge, start to cut the rabbet for the lid.

Take light cuts as vibration can be a big factor. Scraping will surely give you a problem as far as vibration and the possibility of chipping some of the end grain on the joint lines.

Some of the problems with a lid on a round humidor are sealing and locking on (in other words the lid will not come off) due to warping. If we leave enough clearance where warping will not effect it, sealing becomes a problem. Personally I like a recessed lid so I will need clearance around it to assure a removable lid. For sealing, I have chosen a 45 degree sealing surface. In theory, this should seal and if it does warp there is nowhere for it to bind. I have cut my rabbet and started a 45 degree sealing surface.

Progress.

Check the angle often and make sure it is correct. Dismount the barrel but do not remove it from the face plate.

Cutting the 45 degree sealing surface into the corner of the rabbet presents a problem. A straight scrape would probably chatter and possibly dig in, causing Dick to pull his hair out (what's left). To achieve this cut, I have used a square end scraper stood on its edge and used in a sheering cut. Needless-to-say, a light touch and slow movements are necessary. Remember, we have all kinds of joints in this surface between the segments and the cherry and cedar. Work this surface correctly and there should be no need to sand it as we probably would do more harm than good.

Remount screw chuck with lid attached. Start reducing the diameter of the lid to the diameter of the rabbet.

Use care as the lid is composed of segments containing mainly end grain on this surface.

Continue to work the diameter to match the rabbet.

Test fit the lid to the barrel. When removing the lid from the lathe, always remove the chuck with the work piece whenever possible even though a screw chuck has very good repeatability. We always want to keep the work piece concentric to the chuck.

When the lid has a slightly loose fit, remount the chuck and lid in the lathe and cut the 45 degree sealing surface. The width of the angle should be approximately the width of the rabbet's angle.

Now that our angles on the sealing surfaces match, check to make sure that the fit feels correct. If we push down on the lid there should be no sideways movement. It should also be slightly above the top of the barrel so that we can contour the top of the lid.

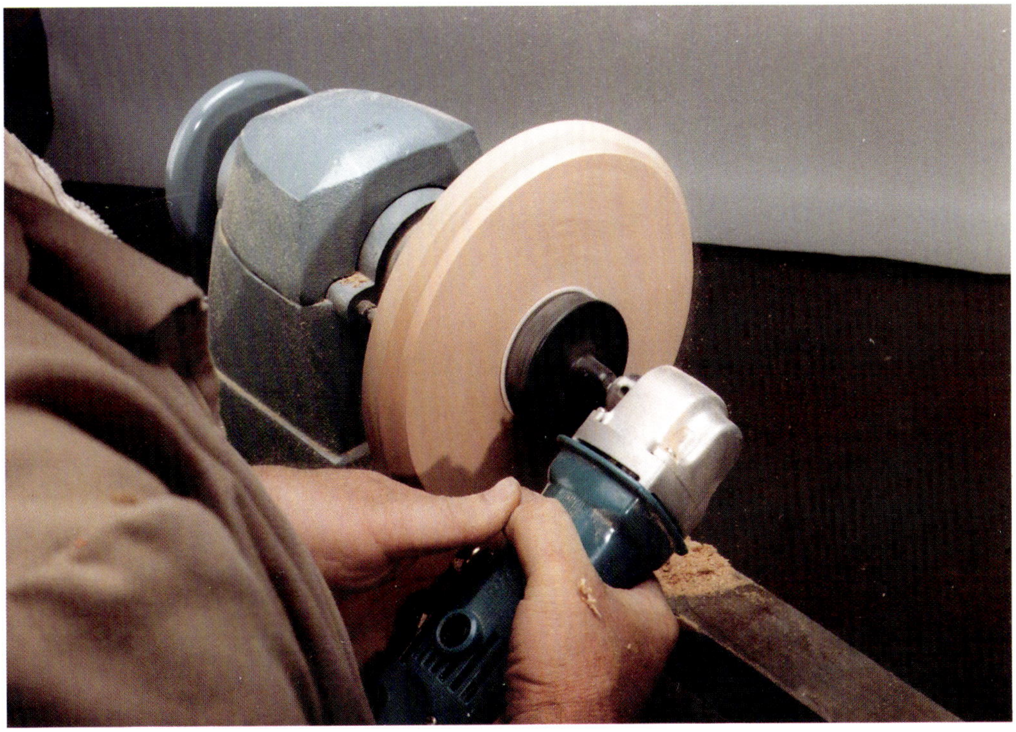

Remount, face off the cedar, and power sand.

Use a sanding block to ensure a flat surface for esthetic reasons.

After power sanding, I always go to hand sanding. I do not believe all the rings can be sanded out by power sanding. I take a sheet of paper and cut it into sixths and fold that into thirds. This gives me a triple thickness sanding pad with many edges to sand up against shoulders. My normal sequence is 180, 220, 320, and 400 grits. I prefer silicon carbide as it is white. On light colored woods, dark paper will loose its grains into the pores of the wood and create a muddy looking finish in light colored woods.

After you are satisfied with the clearance on the straight portion of the lid (not the sealing surface), sand the edge of the lid.

Remount the humidor barrel. Wrap the vertical wall of the lid with enough masking tape to fill up the clearance that we allowed on the lid. This will assure the lid being concentric to the barrel and not rely on our taper although it probably would serve the purpose.

Assemble the lid into the barrel and bring up the tailstock and live center to hold it in place. I use a straight rod tip with a rubber foot on the end for this application.

Start contouring the lid and the top of the barrel.

I prefer to have a crown on the lid instead of a flat surface. It gives something for the eye to flow with.

The lid has a definite flowing crown with a flat spot in the center where we will mount a knob.

On the barrel, again we have flowing curves that compliment the lid. Without these curves, in my opinion, the humidor would look very mechanical. Beauty is in the eye of the beholder.

Power sand, working your way down through the grits as far as you wish.

When you have finished power sanding, sand by hand. I normally run the piece enough to burnish the surface, which shows up all the imperfections. Then I stop the lathe and hand sand them out in the direction of the grain. I work my way through the grits to 400.

Sanded out. Not a scratch is left showing ... but wait until the finish is on. There are sure to be some showing up then.

I use Deft, which is a brushing lacquer. I am brushing on a heavy initial coat of Deft. Allow it to soak for a few minutes.

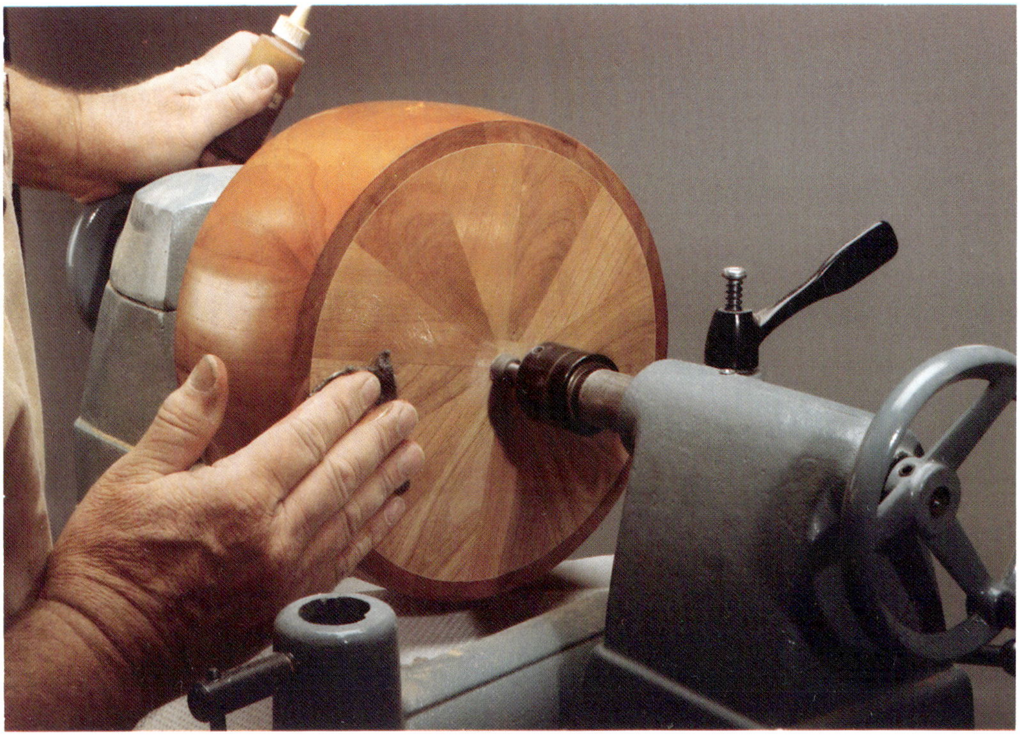

Before it dries, wipe it off with a paper towel and the lathe stopped.

With 0000 steel wool and tung oil finish of your choice, immediately go over the work piece. Naturally, the Deft has not dried and the steel wool and oil combination is removing it from the surface. This makes a slurry that fills the pores yet still gives and oiled finish.

After you have applied a liberal coat of finish with the steel wool, allow it to stand for a little bit to penetrate. I like to cover my lathe to keep it clean. I feel a clean lathe will always produce nicer objects.

After it has penetrated, wipe clean.

Mount a piece of medium density fiber board or some kind of scrap to either a face plate or screw chuck. This will become a jam chuck to reverse the barrel so we can finish the bottom. Work carefully while sizing this because it will be the driving force for the barrel.

The jam chuck is fitted.

We are starting to reduce the excess stock on the bottom of the humidor.

We can reach part of it with the standard tool rest, but after a while it becomes quite a reach across the tool rest which can get us into trouble. If a curved tool rest or holder is available ... now is the time.

A parting of the ways. Using a parting tool, work side to side to provide enough clearance that we do not bind up the tool. Keep reducing the diameter of the bottom of the barrel. Work carefully. We are too close to the end to lose it now!

The piece has been removed. This surface now needs to be cleaned up as it will be the bottom of the humidor.

Sometimes it is better to use your head than your heart. At this point I was getting close to jamming up my parting tool and was reaching an awfully long distance over the tool rest. At a time like this, a hand saw works wonders.

Using duct tape as a safety factor.

Take a 3/8 spindle gouge and start facing off the bottom of the humidor.

The bottom of the humidor should be slightly concave. If it is flat, it will never sit evenly on anything. This way it will sit on the rim.

Being a wood turner, I always like to leave my mark in a normally hard to reach spot, such as a base. In other words, I like to let people know I've been there! A small bead will accomplish it and also give the base just a touch of class.

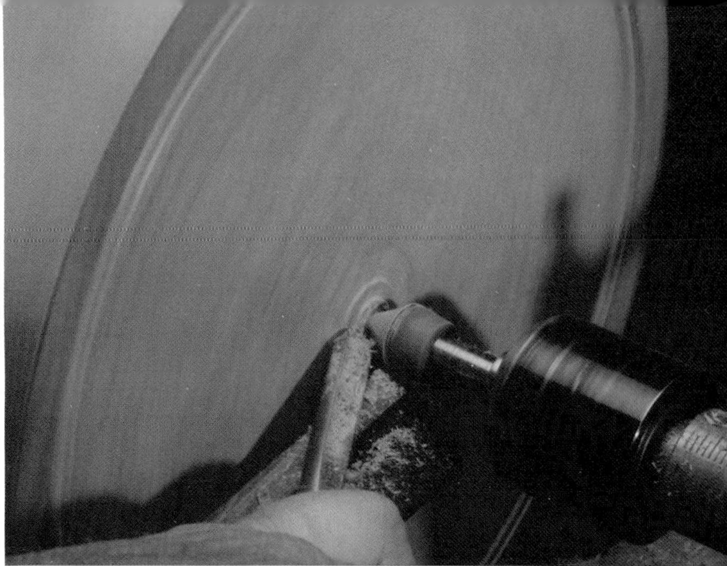

We have to reduce the last spigot where the live center is applying pressure. Once it is gone, remember that the only thing holding the humidor is the jam chuck fit and the insurance of the tape. Take light clean cuts.

The bottom is ready for sanding.

Power sanding.

After you have power sanded, hand sand the base, working your way through the grits.

The sanded base, ready for finish.

The finish is applied. Let it set for a few minutes to absorb.

Applying a generous coat of Deft finish to the bottom.

Wipe off the excess finish with a paper towel.

Immediately we are going to apply the tung oil finish with 0000 steel wool. Wipe off the excess again.

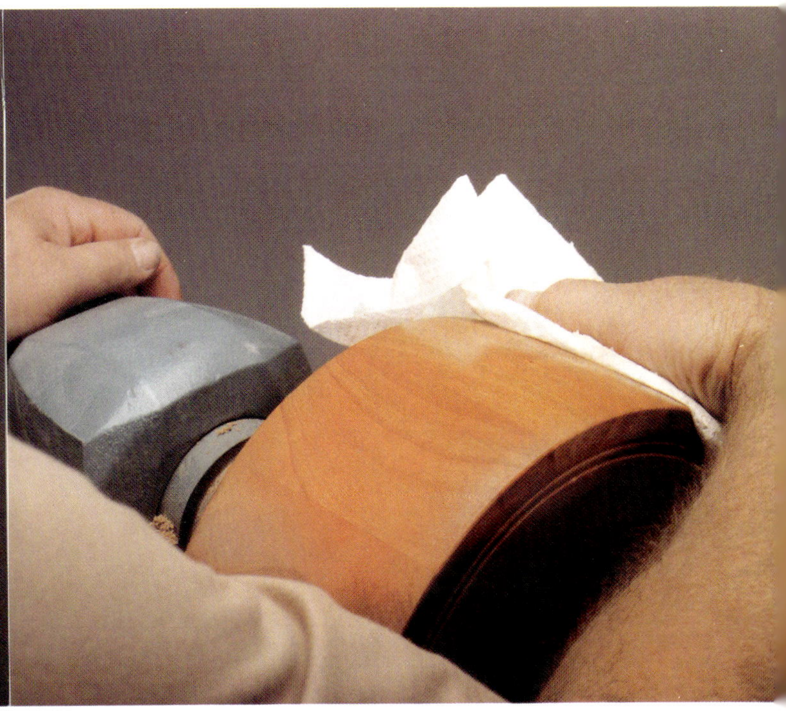

Ready to remove the humidor from the jam chuck. If there is any pick up from the tape, now is an excellent time to remove it with the steel wool and tung oil while it is still held by the jam chuck.

The handle. We are going to start with a piece of cherry and a scroll chuck. If this chuck is not available, a scrap block and double face tape would work adequately.

Be careful that you do not contact the jaws with your tools as it creates a disturbing noise and it is very rough on your equipment.

Reducing half of the square to round (on account of the jaws) so that we can reverse the piece.

We have reversed the cherry block and are ready to clean up the other half.

Face off what will be our mating surface to the lid and start shaping our knob.

Remove the drill chuck from the tailstock and mount it in the headstock. I have inserted a metal 1/4" dowel into the chuck. This will be the drive to finish our knob.

Put a drill chuck with a 1/4" drill into the tailstock. Bring up to the rotating work piece and allow the drill to seek its center. Tighten down the tailstock and drill to a half inch depth. This hole will hold a dowel which will be glued to the lid. Make sure that surface is flat or very slightly concave so that the outside edges of our mating surface make a tight match.

Put the knob on the dowel and bring up the live center. To help create friction to drive this, make sure the bottom of the knob is against the jaws of the chuck. Bring up the tailstock and apply pressure. Start shaping your piece. If you apply too much pressure with the tool, you will immediately realize there is not much driving it.

The knob has been shaped and sanded.

A coat of Deft and oil with 0000 steel wool has been applied. It is ready to be mounted on the lid.

Cut a piece of 1/4" dowel slightly shorter than the combined depths of both holes and glue together.

Now that we have our humidor constructed and finished, we need to add a hygrometer, humidistat, and some dividers. The hygrometer measures the humidity inside of the humidor, the humidistat provides the humidity, and the dividers separate the cigars to maintain each brand's distinctive taste. The humidistat is going to be attached to the lid where we could also attach the hygrometer. In this location you would have to tip the lid upside down to read the gauge and there is some chance you would damage the gauge every time you set down the lid. I choose to combine dividers and a provision for the hygrometer in one unit. I have chosen 3/16" thickness, 2 3/8" wide Spanish Cedar for the dividers and a piece of 7/16", 3" x 2 5/8" in which to mount the hygrometer. This piece will also be used to space the dividers. These dimensions are not etched in stone. Use whatever appeals to your needs or desires for your particular humidor.

We need to scribe the arc of the inside of the humidor onto the mounting block. Position the block against the wall and use a short pencil parallel to the wall to scribe the radius onto the block.

The line is scribed on the block.

After cutting off the excess, go to the sander and fit the block to the inside diameter of the humidor.

To find the center of our mounting block, go corner to corner both directions. The radius on one side creates an off-center mark. Measure from both sides to the mark and split the difference. The hole will now be centered to the block.

The hygrometer requires a 1 3/8" diameter hole. I used a Fostner bit in my drill press. This bit cuts an exceptionally clean, precise hole which is necessary to hold the hygrometer securely.

Dry fit your pieces prior to gluing.

Glue the dividers on either side of the hygrometer mounting block with a slight protrusion past the radius. When dry, go back to the sander and blend the dividers to the radius on the block.

I have left one end free. The purpose of this is that we can squeeze the free end slightly and insert or remove the dividers. Use the sander to match the contour on the ends of the dividers, working slowly until you reach the fit you desire.

Insert the hygrometer and the assembly is complete.

Attaching the humidistat is one of the more complicated steps in this humidor. It consists of attaching a piece of self sticking velcro to the lid and the opposing to the humidistat. I think I can even handle this!

Take a piece of tape and use it to clean the dust from the area where you intend to attach the velcro strip. This will help it to adhere much better.

The humidistat is in place.

The humidor is complete.

# Gallery

# A Final Word

To those of you who were paying close attention and were very observant to the changes and inconsistency of the grain patterns in the wood, yes, there is a reason for it. Like a great chef preparing a culinary work of art for the media, Dick has taken liberties. Not wanting to spend time waiting for the "glue" to dry during the photo shoot, some of the pieces were preglued, allowed to dry, turned, and substituted for their drying counterparts shown in the turning and assembly instructions. The results are the same. A humidor worthy of an Aficionado. *Cindy Sing ... Wife.*